AF228410

Searchlight BOOKS™

Hunting and Fishing

Freshwater Fishing

Diane Lindsey Reeves

Lerner Publications ◆ Minneapolis

Copyright © 2024 by Lerner Publishing Group, Inc.

All rights reserved. International copyright secured. No part of this book may be reproduced, stored in a retrieval system, or transmitted in any form or by any means—electronic, mechanical, photocopying, recording, or otherwise—without the prior written permission of Lerner Publishing Group, Inc., except for the inclusion of brief quotations in an acknowledged review.

Lerner Publications Company
An imprint of Lerner Publishing Group, Inc.
241 First Avenue North
Minneapolis, MN 55401 USA

For reading levels and more information, look up this title at www.lernerbooks.com.

Main body text set in Adrianna Regular.
Typeface provided by Chank.

Library of Congress Cataloging-in-Publication Data

Names: Reeves, Diane Lindsey, 1959–author.
Title: Freshwater fishing / Diane Lindsey Reeves.
Description: Minneapolis : Lerner Publications, [2024] | Series: Searchlight books - hunting and fishing | Includes bibliographical references and index. | Audience: Ages 8–11 | Audience: Grades 4–6 | Summary: "Freshwater fishing is a popular outdoor sport. People fish on lakes, rivers, and other sources of freshwater. Learn everything fishers need to know about freshwater fishing, including the necessary equipment and taking care of nature"—Provided by publisher.
Identifiers: LCCN 2022042945 (print) | LCCN 2022042946 (ebook) | ISBN 9781728491578 (library binding) | ISBN 9798765603765 (paperback) | ISBN 9798765600474 (ebook)
Subjects: LCSH: Fishing—Juvenile literature. | Freshwater fishes—Juvenile literature.
Classification: LCC SH445 .R44 2024 (print) | LCC SH445 (ebook) | DDC 597/.48—dc23/eng/20230111

LC record available at https://lccn.loc.gov/2022042945
LC ebook record available at https://lccn.loc.gov/2022042946

Manufactured in the United States of America
3-1010680-51106-2/20/2024

Table of Contents

Chapter 1

GONE FISHING

A new day is dawning. The sun is peeking through the clouds. The weather is pleasant and cool. The water is calm and clear. All around nature is waking up. Birds are tweeting. Fish are on the move looking for food. Quick! Grab your fishing pole. It's time to go fishing!

An Equal Opportunity Sport

Fishing is one of the most popular outdoor sports in the US. More people fish than play golf and tennis combined. It's a sport that any age or gender can master. You can do it alone or with family and friends. The thrill of the chase keeps fishers, or anglers, coming back for more.

Where and What to Fish

Freshwater fishing happens on lakes, reservoirs, ponds, streams, and rivers. Some fishers fish from the bank. They cast off from the shore, a bridge, or a pier. Other fishers head out in boats. Anything from a simple canoe or jon boat to a fancy pontoon or fishing boat will do the job.

Some of the most common freshwater fish include different types of bass and catfish. Crappie, trout, and walleye are also plentiful in the US.

Gearing Up to Fish

Basic gear needed to fish includes a rod, a reel, and a well-stocked tackle box. A fishing rod is a long thin pole which has a line and hook attached to it. A reel is attached to a fishing rod and is used to wind and unwind fishing line. Tackle includes all the gear needed to fish including bait, hooks, and line.

Rods and Reels

There are four main types of fishing rods and reels: spin-casting, spinning, bait casting, and fly rods. Different types of rods are best for specific types of fishing.

Spin-casting rod and reel combinations are the easiest to use. You cast off by pushing a button with your thumb. Beginners should use a rod that is no longer than their height for better control.

Spinning reels are used by anglers ready for more challenge. Experienced fishers control the line with their fingers. This gives them more control to cast farther and more accurately.

Baitcasting reels are used to catch larger fish like bass. Once a fisher gets the hang of using the thumb control, they can cast off with precision.

Fly Fishing

Fly rods and reels are used with lightweight lures that look like an insect such as a fly to catch fish. The fisher stands in the water and whips the rod back and forth over their shoulder. It is a relaxing way to fish for trout and other fish.

FLY FISHERS USE LURES THAT LOOK LIKE FLIES TO CATCH FISH.

Fishing History

In ancient times fishing was a means of survival, not a sport. Fishers back then didn't have the fancy gear used today. Instead, they used sharpened sticks called spears to hunt for food. They also used barbed spears called harpoons to hunt for whales and large fish like swordfish. Art found in caves and historical records dating back as far as 16,000 years refer to early fishing practices.

This sculpture honors fishing traditions in Martha's Vineyard off the coast of Massachusetts.

Tackling the Tackle Box

A tackle box or bag has compartments and drawers for keeping fishing gear organized. A well-stocked tackle box includes extra line, good quality hooks in different sizes, needle nose pliers, and a small first aid kit. It also includes live bait, bobbers, sinkers, and lures.

Live bait like nightcrawlers is attached to hooks to attract hungry fish. Bobbers or floats keep bait floating at a depth where fish are feeding. They bob around when a fish bites. Sinkers are weights fishers use to help sink the bait down into the water. Non-toxic, lead-free sinkers are the most environmentally friendly option. Lures are small, colorful objects often shaped like a fish's prey. The bright colors lure fish to your line.

Chapter 2

SAFE AND SOUND

Paying attention to weather is the most important way freshwater fishers stay safe. Some fishers enjoy fishing in a light drizzle. But fishing in a rainstorm is dangerous. It is never okay to be out on the water during a thunderstorm. High winds and rough water are best to avoid as well. Always check—and double check—weather reports before heading out.

Wearing life jackets while on a boat in the water is also a good idea. This is true even for good swimmers. This added layer of protection helps if someone falls overboard or gets injured.

Hooked on Safety

Freshwater fishers know they are more likely to hook fish with strong, sharp hooks. They also know to be careful when using them. It is possible to hook a friend or family member by accident when casting a line. That's why it is important to keep track of where fellow fishers are standing and keep your lines and sharp hooks away.

Sharp knives are used to cut lines and disentangle fish from hooks. Fishers keep a first aid kit in their tackle boxes in case of a cut or other injury.

A SHARP KNIFE COMES IN HANDY FOR CUTTING FISHING LINES.

Cool and Comfortable

Fishers prefer to avoid sunburn and bugs when out on the water. They should apply sunscreen with at least 30 SPF protection to all exposed areas of skin—even on cloudy days! Fishers also need to protect their eyes from harmful rays from the sun by wearing sunglasses and a hat.

For extra protection, you may want to use sun
protective clothing. Sunscreen clothes are made from
special fabric designed to reflect light and absorb
ultraviolet rays. You can find sunscreen shirts, jackets, and
pants at any sporting goods store.

Insect repellent comes in handy for buggy situations.
This is available in spray cans you apply to exposed skin
or handy wristbands you slip on your wrists.

Chapter 3

THINGS FISHERS SHOULD KNOW

Freshwater fishing is the most popular type of fishing in the US. The sport attracts more than 40 million anglers every year. Other types, like deep-sea fishing, saltwater fishing, fly fishing, and ice fishing, have many fans as well. Commercial fishing is a huge industry around the world. In the US alone, commercial fishing pulls in over 9 billion pounds of seafood a year.

Fishing History

People from all walks of life enjoy fishing. Several US presidents were known to spend time on the water with a rod and bait. This includes George Washington, Grover Cleveland, Herbert Hoover, Franklin D. Roosevelt, Harry Truman, Dwight Eisenhower, Jimmy Carter, George Bush, and George W. Bush.

This statue of President Herbert Hoover shows him enjoying a favorite pastime.

Favorite Fishing Holes

Wherever you find freshwater, you are likely to find fish. Fishers take motorboats out on big natural lakes, rivers, and human-made reservoirs. They also paddle through quiet creeks, ponds, and streams. One of the best things about freshwater fishing is the variety. You can find out about local fishing spots through your state fish and wildlife agency. Or better yet—ask other fishers.

In most states, fishers over the age of 16 need to purchase a fishing license. These can be easily obtained online or at local sporting goods stores.

Fishing Season Is In

Freshwater fishing is nearly a year-round sport. Different times of day are best for different seasons. Spring and fall fishing are best in the afternoon to early evening. In summer it's better to avoid hot afternoons and fish at dawn or dusk. Winter fishing is possible in places with warmer year-round climates.

STEM Spotlight

Gills are to fish what lungs are to humans. Instead of getting oxygen from air, fish get it from water. They take water in by opening and closing their lips. The water filters through the gills with rows of feathery filaments made of protein molecules. These filaments have thousands of tiny blood vessels that move the oxygen into the bloodstream. Fish gills even have more blood vessels than human lungs!

Fish breathe in oxygen contained in water.

Chapter 4

TAKING CARE OF NATURE

People who enjoy outdoor sports have a responsibility to take care of nature. Buying fishing licenses is one way that fishers do this. These fees are used to fund conservation efforts. Conservation programs protect wildlife and natural resources. They make sure that there are plenty of fish now and in the future.

There are other simple ways that fishers can do their part. One is to never leave litter in waterways. Even better—pick up litter found out on the water. This helps keep waterways free of plastics and other debris that harm fish.

Overfishing depletes fish supplies and upsets the balance of ecosystems. Limiting the number of fish caught helps conserve nature.

Catch and release practices protect fish populations.

Catch and Release

Sometimes people fish to catch food to feed their families. That is fair and ethical.

Other times fishers use a catch and release method of fishing. That means after they catch a fish, they quickly and carefully remove the hook. Then they set the fish free in the water to swim away. This is another helpful way to conserve nature.

Fishing Tips

- Do online research or check out a fishing book at the library to learn about good fishing spots.

- Plan ahead to bring the best gear for the type of fish you are after.

- Use sharp hooks and strong lines.

- Check the weather forecasts and dress for the weather.

- Make sure someone knows where you are going and when you expect to return.

- Wear life jackets anytime you are in a boat.

Glossary

angler: a person who fishes with a rod, line, and hook

cast: the act of throwing bait or a lure using a fishing rod, reel, and line

conservation: to prevent wasteful use of a natural resource

ethical: following accepted rules for right and wrong behavior

jon boat: a small, lightweight fishing boat with a flat or nearly flat bottom

pontoon: a flattish boat that relies on floats to remain buoyant

reel: a device for winding and unwinding fishing line

rod: a long thin pole with a line and hook attached to it used for catching fish

tackle: all the gear used in fishing including hooks, lines, bait, sinkers, and floats

Learn More

Carpenter, Tom. *Fishing*. Minneapolis: Sportszone, 2020.

Doyle, Abby Badach. *Freshwater Fishing*. NY: Gareth Stevens Publishing, 2023.

Kiddle: Fishing Facts for Kids
https://kids.kiddle.co/Fishing

Kingston, Seth. *Fishing*. NY: Powerkids, 2022.

National Geographic Kids: Fish
https://kids.nationalgeographic.com/animals/fish

National Park Service: Fish & Fishing
https://www.nps.gov/subjects/fishing/index.htm

Index

Photo Acknowledgments

Image Credits: p. 5; Lightfield Studios/Shutterstock, p. 6; Rocksweeper/Shutterstock, p. 7; dasytnik/Shutterstock, p. 8; Afanasiev Andrii/Shutterstock, p. 9; Dimitriev Mikhail/Shutterstock, p. 10; Annette Shaff, p. 11; Wirestock/Dreamstock, p. 12; SnapTPhotography/Shutterstock, p. 13; alexnika/Shutterstock, p. 15; Daniel Thornberg/Dreamstock, p. 16; AT Productions/Shutterstock, p. 17; Zadorozhnyi Voktor/Shutterstock, p. 18; Thomson_1/Shutterstock; p. 19; Adam 121/Dreamstime, p. 21; Joe Sohm/Dreamstime, p. 22; Daniel Thornberg/Dreamstime, p. 23; Anne Beruldsen/Dreamstime, p. 24; Twildlife/Dreamstime, p. 25; pfluegler-photo/Shutterstock, p. 27; itakdalee/Shutterstock, p. 28; CameraLens Pro/Shutterstock.

Cover: Terry Vine/Getty Images.